Victoria & Albert Museum

Lighting

The Arts and Living

Alastair Laing

London: Her Majesty's Stationery Office

First published 1982

Acknowledgements
My first thanks are due to the original general editors of this series, John Fleming and Hugh Honour, both for their invitation to write this little book, and for their patient encouragement of a laborious author. From another writer in this series, Eileen Harris, I received inspiration, help, and good advice. I should also like to thank those members of departments in the Museum who helped me in the search for apt objects to illustrate, and those in the National Art Library and London Library who helped me to track down books. Amongst the friends who alerted me to references to lighting of which I was unaware, I should particularly like to remember the late Hugh Murray Baillie, with deep regret that he did not live to see this book appear.

With the following exceptions all the objects illustrated are from the Victoria and Albert Museum. Fig 1 British Museum; Fig 7 Louvre, Paris; Fig 10 Albertina, Vienna; Fig 16 Musée la Tour, Saint Quenten; Fig 24 The British Library.
Fig 25 reproduced by courtesy of the National Monuments Record and Plate 9 reproduced by courtesy of Barnaby's Picture Library.

Design by HMSO Graphic Design

Cover. A silver lamp in the form of Cupid, by Giovacchino Belli, *c.*1805. The butterflies act as shades and the quiver contains tweezers and other equipment for tending the wick.

ISBN 0 11 290284 7

Contents

1 Introduction

'When light was brought in the evening, the Ancient Greeks used to say "Hail! friendly light". Instead of this expression, the Greeks of today say "Good Evening" to one another when there is company. The same thing happens in Italy, in Provence, and elsewhere. In Paris nothing is said.' With this observation the seventeenth-century French writer Ménage registered a profound revolution in attitudes toward light. Until continuous advances in technology fostered a manipulative rather than an awestruck attitude towards Nature, light was universally regarded as having not just a utilitarian, but also a religious significance. The salutation recorded by Ménage is still to be found with the 'Namaste' of the Hindu, and at the lighting of the Sabbath lights by the mother of a Jewish household; most religions have had their gods of light, sun and fire, and in dualistic systems Light is set off from Darkness, as Good from Evil. The Bible traces Creation to God's declaration 'Let there be Light', whilst Żoroaster went further, and identified God with light – Ahura Mazda. This identification of God as 'The Light of the World' was transmitted both to Christians – whence Holman Hunt's famous picture – and Muslims – whence the superb Syrian and Egyptian enamelled glass hanging mosque lamp bowls of the thirteenth to fifteenth centuries, whose usual inscription from the Koran declares their symbolic purpose: 'God is the light of the heavens and the earth; his light is like a niche in which is a lamp, the lamp in glass and the glass like a brilliant star.' [*pl.1*].

Lighting, or the fire that was once needed to engender it, was thus commonly regarded as either a beneficent gift of the gods, or the result of a liberating theft from them. Our remoteness from the days when some form of flame was essential to the provision of artificial light makes it easy to forget that, when Prometheus smuggled fire from Olympus in a fennel-stalk, he was bringing to mankind the possibility of civilization, by offering the choice not merely between the dualities of cold and warm, raw and cooked, but also between light and darkness.

The story of lighting is more than a record of the changes brought about by improvements in its technology. The increasing security and diversity of human existence made possible by the spread and increased sophistication of lighting bind it intimately to the history of civilization. Three broad divisions in the evolving relationship of man to lighting may be discerned: in the first, after the initial domestication of the naked flame for the purposes of lighting had been achieved, there was not much further development in either the techniques or the quantity of available lighting; lavish lighting was primarily religious in significance; and its domestic use remained an exceptional, and mostly functional, prolongation of the hours of daylight. In the second phase, the accumulation of surplus wealth on the part of the possessing classes was turned towards the increasing consumption of light, both for display and to facilitate modes and hours of existence impossible to those who did not possess such means, so that lighting became an important factor in social differentiation; this was only possible because of the kind of secularization in attitudes toward lighting noted by Ménage, though religious pretexts for its use persisted. In the last era, in which we ourselves live, technical advances and the spread of wealth have led to the steady diminution of social distinctions in the use of light, and to its general availability. As a result, light is nowadays taken for granted; the basic human necessities are defined as 'warmth, food and shelter', and only a blackout is capable of bringing home to us the extent to which our very patterns of existence are dependent upon light being at our beck and call. The inadequacies of candles, for over a millenium the essential form of lighting in Northern Europe, but now just used in church or to create atmosphere, quickly emerge. Otherwise, we are mostly oblivious of the extent to which the transformation of artificial light from a luxury to a utility, from an inconstant, impermanent and precarious flame to a steady glare, have transformed the way we both look upon and experience our surroundings, removing whole areas of aesthetic perception and wonder, diminishing great tracts of fear, altering our domestic environment, and sapping our awareness of season and the passing of time.

Light within and light without

The history of artificial lighting is related to that of the manipulation of natural light through windows. The modern office block with glass walls lit internally by fluorescent strips throughout the brightest summer's day reveals, however, that the relationship is not quite as straightforward as might be expected. Initially the spread of glazing and the increase in the size of windows diminished the need for artificial lighting in the daytime. By the elimination of draughts, glazing also improved the insulation of rooms, in which candles could burn slowly and regularly, encouraging their occupants to stay up after dark. But the increase of indoor light from outside in the daytime has reduced tolerance of dimness and shadow. In the eighteenth century some eyes were so sensitive that they had to be shielded from the light of a single candle. And, in the early nineteenth century, Mme de Genlis deplored the introduction of bright lamps which were replacing candles and would, she asserted, ruin the eyes of the young. Yet we are constantly admonished not to strain our eyes through reading by a light of less than forty foot-candles (i.e. at a foot away, forty times the light of a standard candle) and find any kind of shadow or flicker intolerable for anything save candlelit suppers. It is, of course, true that the elimination of dark areas diminishes all kinds of hazards, but the constant level of artificial lighting in modern office blocks is the visual counterpart of the psychologically calculated use of Musak. The approaching energy crisis threatens not only our mobility, but also our unthinking dependence upon the ubiquitous availability of artificial light. The almost complete absence of studies of the social history of lighting (with the striking exception of O'Dea's) shows how it has been taken for granted. In his five-volume *Illustrated English Social History*, G. M. Trevelyan devoted one line to lighting, that recording the arrival of gas and electricity; the aim of this booklet is to show that there is more to the history of lighting than a mere record of technical improvements and variations.

2 Lighting in Antiquity and in the Early Church

Early lighting devices

The first artificial lighting was firelight. The fire by which men kept warm and on which they cooked also provided light to see by. Brands plucked from the fire made torches to carry, and just as resinous wood was found to flame more vigorously, so it must soon have been observed that the fat exuded by roasting animals made fires burn more brightly. From there it was easy to lift a fire up in a metal holder to enable it to cast a wider gleam – whence the brazier and the fire-bucket or cresset (in its original sense). And it only required the collection of fat or oil in a container with a piece of inflammable material as wick to produce the lamp.

Since anything that will burn will yield some form of light, whatever is ready to hand and cheap has been so used: all kinds of tallow have been extracted, from the stinking bullock-fat imported into England in the sixteenth century as a consequence of the trading-agreement with Muscovy, to the vegetable tallow procured in China from the 'tallow tree' (*Stillingia sebifera*); oils were expressed from everything from fish to the seeds of almost anything from grapes to tea; wax came not only from bees, but, shaded a pleasing green and less liable to wilt, from the American bayberry; whale blubber was a cheap source of supply in the nineteenth century, while the head cavity of the sperm whale yielded abundant spermaceti, which, Thomas Browne discovered, 'flameth white and candent like camphire [itself a vegetable oil from an aromatic shrub] . . . some lumps . . . afford a fresh and flosculous smell'; untreated carcases have even been used – from the dogfish tails and candlefish stuck into a cleft stick, used by Newfoundland fishermen and American Indians respectively, to the stormy petrels with wicks thrust down their throats burnt in the Shetlands. British soldiers at Ypres in the First World War improvised lamps out of sardine-tins using rancid oil from the sardines. Most macabre of all was the use of the fatty matter called *adipocire*, which was excavated from the cemetery of the Innocents in Paris after its suppression at the end of the eighteenth century.

From the invention of the lamp until the end of the eighteenth century there was virtually no improvement in the techniques or efficiency of lighting, but only unceasing refinement and embellishment of its vessels; the only major technical discovery was the candle, which, being of solid fuel, was more portable, and, requiring supports rather than containers, allowed a greater variety of forms and materials in their manufacture.

The earliest and most basic form of oil lamp is the hollow stone. Over one hundred of these have been found in the caves of Lascaux, and by their light the cave-paintings must have been done, over 12 000 years ago. A sandstone lamp of the same Magdalenian period also found in a cave in the Dordogne already has the first essential refinement, a spout, so that the wick can be held, and its consumption of oil controlled by the extent of its protrusion; the more of the wick that shows, the brighter it burns – but also the smokier. Despite, or indeed, because of their crudity, stone lamps continued in use as the most basic form of lighting in our own civilization into the Middle Ages – in the form of the cresset stones used to light church porches and the cloisters of monasteries, and amongst the Eskimos, using blubber for fuel and moss wicks, down to our own day. At the dawn of civilization in Mesopotamia around 4000 B.C. shells were used, which had the advantage of natural spouts. These shells provided the models for some of the earliest fabricated lamps (probably burning sesame-seed oil), such as those in the form of bifurcated conches found at Ur, and the most primitive form of pottery lamp, the Phoenician terracotta 'cocked-hat' lamps inspired by the scallop. Gold, silver, copper, and alabaster were used for the lamps that were found in the royal tombs of Ur – revealing the non-utilitarian uses of lighting from the start. The Ancient Egyptians appear only to have used undistinctive saucers with floating wicks for lamps, but these might be placed on yard-high stands decorated with hieroglyphs, or inside a transparent painted alabaster lamp-shade like the one found in Tutankhamen's tomb.

The vast quantities of oil storage jars found in Mycenaean and Minoan palaces are a pointer to the number of lamps used to light their labyrinthine recesses, but the collapse of this civilization resulted in a reversion to resinous wooden torches, often held aloft by human torch-bearers, as Homer reveals. In the early seventh century B.C. Athens began to manufacture 'cocked-hat' lamps, and, progressively adding improvements in design such as the bridged nozzle (allowing only the end of the wick to appear, thus minimizing smoke), handles, folding over the rim (to reduce spillage), and glazing to counter porosity, developed a major export trade in terracotta lamps. As time went on, the top was steadily closed over to protect the oil, leaving no more than a hole to replenish the lamp by, and creating a disc inviting decoration – which became easy with the switch from wheel-throwing to

Figure 1. A typical Roman terracotta oil-lamp, 1st century A.D. The disc is stamped with an allegorical figure of Abundance.

moulding as the method of manufacture at the beginning of the third century B.C. The Romans, who took over the mass manufacture of lamps from the Greeks, decorated these discs with everything from deities to obscenities, and also made terracotta lamps in a variety of human and animal shapes [*fig.1*]. The more sophisticated of these were adaptations of bronze lamps, in the form of satyrs, ships, and feet – which were to inspire exquisite imitations in the Renaissance, above all in Padua [*fig.2*]. Though the use of gold is recorded for temple lamps, bronze was the normal material for luxury lamps and such things as the multiple-nozzled hanging-lamps (*lychnuchi*) and lamp-stands (*candelabra*, so-called because their design derived from the candle-holders of the Etruscans), whose prodigality of light only the rich could afford.

Yet olive oil was sufficiently plentiful (the very word 'oil' derives from the word 'olive') and lamps sufficiently economical in its consumption (a single-spouted lamp could give fifty hours light), for some form of lighting to have been within most people's reach: Herodotus's expression for evening was 'lamp lighting-up time'. But Varro and Martial

Figure 2. A Renaissance pseudo-Antique bronze oil-lamp in the shape of a boat, made in Padua by Riccio, *c.*1500.

reveal that, whereas the Roman rich adopted the Greek lamp *(lychnus* or *lucerna)*, the Roman poor stuck to the Etruscan *candela* – in their case more often a piece of tow steeped in tallow, or a rush-light or taper, than a true candle of wax *(cereus)*, which was something more special. Lamp wicks were commonly of mulleine *(verbascum)*, but also of flax, hemp, and even, in the case of the perpetual flame tended by the Vestal Virgins, asbestos. Enlarging the wick promoted more smoke than light, so nozzles were multiplied instead, as many as sixteen to one lamp. When light was lavishly consumed it was impossible to keep all the wicks in trim, so that slaves had to wash down the decorations and statues after banquets. Vitruvius counselled against frescoes and stucco reliefs in winter apartments because of lamp-black, recommending polished black walls with ochre or vermilion panels instead. Lamps smoke worst just before dying, but superstition prevented the Romans from ever extinguishing lamps once lit. Though this was rationalized as an encouragement to hospitality, it more probably reflected a primitive reverence for light itself, like the lamps kept perpetually burning before their household gods.

The classical day was governed by the hours of daylight, so the use of lamps far into the night was the result of either study (whence the word 'lucubration') or the pursuit of pleasure. It was customary to get up at or even before dawn, in order to profit by every daylight hour, whilst the evening *cena* was eaten, or at the very least begun, well before dusk. The banquets of the rich went on to all hours, for only they, in the absence of any street-lighting, had slaves with torches to guide them safely home.

Lighting and the Early Church

After the third century A.D. there was an evident decline in lamp manufacture in the north-west provinces of the Roman Empire; disturbed trade and communications may have made oil scarce – but it also seems that candles increasingly supplied the deficiency. After the fall of the Empire this substitution was encouraged by the Western Church, which, having lost whole tracts of the olive- and wine-growing world, fostered bee-keeping to produce wax and planted new northern vineyards for wine, in order to provide for its liturgical needs.

The church was initially hostile to 'lighting useless candles at noonday' (Tertullian), because of the employment of lights in the ceremonies of the pagans and the Jews: as Lactantius said around A.D. 300: 'Surely he cannot be reckoned sane who offers as a gift to the author and giver of light the light of candles and tapers. Very different is the light he demands from us, not smoky, but as the poet says, pure and clear, that is, of the mind.' But as Christianity turned into the state religion, churches became the recipients of gifts of hanging lamps and candlesticks in precious metals with hundreds of lights; and already around A.D. 400 'the bright altars are crowned with thronging lamps . . . day and night they burn' (Paulinus of Nola). The form of the earliest candlesticks – with prickets, large drip-pans, knopped shafts and tripod feet *[pl.2]* – was simply adapted from that of lamp-stands, and already established by the time of Justinian. By the First Council of Braga in 561 it was ordered that the revenues of a church should be divided into three parts: a third for the priest, a third for the bishop, and a third for the repair and lighting of the church. Early Christian Fathers celebrated their churches as *splendens* and *micans*, resplendent and glittering, a far cry from Milton's 'dim, religious light.' Enormous candelabra, cross-shaped chandeliers (one in St Peter's had 1365 lights) and *coronae* are recorded in the eighth century – the precursor of those great twelve-turretted Romanesque wheels (as at Aachen and Hildesheim) that symbolized the Heavenly Jerusalem. A remark made by Epiphanius (around A.D. 400) shows that the church might be the only building showing a light in a town at night. Not until the eleventh century, however, were candles placed upon the altar; at first only a pair, but in the thirteenth century their multiplication to six or seven for High Masses

Figure 3. Candlestick-bearing angels symbolic of acolytes, carved in limewood by Tilman Riemenschneider, *c*.1500.

began (with the Cistercians fighting a rearguard action both against the use of wax for candles and against their proliferation, as they had against stained glass). Sculpted angel-candelabra evoke the earlier habit of the requisite candles for Mass being held by acolytes rather than being placed upon the altar [*fig.3*]. In the late Middle Ages bequests of candlesticks and endowment of sanctuary lights became one of the commonest provisions of wills. The institution by Pope Zosimus in 417–18 of a specially large candle to stay lit from Easter Sunday to Ascension Day gave rise to the most massive and ornate furnishing of major churches – the Paschal Candelabrum; one of the hugest was that of Durham Cathedral, whose branches stretched from one wall of the choir to the other. The Feast of the Presentation came to be set aside as the day on which all candles were blessed – Candlemas – because that was the day on which Christ, the true 'light to lighten the Gentiles' was introduced into the Temple. Justifying this by confused symbolism connected with the supposedly virgin bee, the Church began to insist on the exclusive use of wax in its candles from the thirteenth century onwards. Metal has always been the usual material for candlesticks, because it neither breaks nor burns; though the church used precious metals for the most special candlesticks *[fig.4]*, iron and brass – which does not oxidize – were the norm. A whole export industry in these and other liturgical objects in brass grew up from the valleys of the Meuse and the Elbe, and in enamel from Limoges *[pl.2]*.

Figure 4.
The 'Gloucester Candlestick', gilt bell-metal, probably English, *c.*1110. This piece was originally made for the abbey church of St Peter's, Gloucester.

3 Domestic Lighting

The Early Middle Ages

Meanwhile, during the Dark and Middle Ages, domestic lighting remained crude and lagged behind that of churches (even Late Medieval glossaries still list candles amongst 'things pertaining to the church'). One of the earliest known candlesticks, found in an Alemannic chieftain's grave of the sixth/seventh century A.D., is merely of turned wood. Gregory of Tours records a reversion to barbaric notions of splendour, with human torch-bearers at banquets. These were the models for the candlesticks shaped like standing men and horsemen also produced by the Mosan brass-founders [*fig.5*], which were probably for secular rather than ecclesiastical use (though the earliest specific mention of '*Chandeliers à mettre à table*' is not until 1328), and for the

Figure 5. A Mosan bronze candlestick in the form of a centaur, probably designed for household use, 13th century.

Plate 1. A hanging mosque lamp of enamelled glass with wire cords. Damascus, *c.*1340.

Plate 2. A Limoges *champlevé* enamel tripod pricket candlestick, 13th century.

Plate 3. A German stained-glass window, showing Tobias and Sarah in bed, with the candlestick on a *buffet* at its foot, 16th century.

Plate 4. Three examples, in a variety of materials, showing the assimilation in the West of the socketed candlestick from Persia. (a *left*) historiated Urbino maiolica, *c.*1535; (b *centre*) early 16th century Veneto-Saracenic brass, inlaid with silver and niello, early 16th century; (c *right*) Venetian reticello glass, 16th century.

Plate 5. Candlesticks in differing materials: (a *left*) Limoges enamel, by Pierre Reymond, French, 1556; (b *centre*) pewter, the 'Grainger Candlestick', English, 1616; (c *right*) earthenware, Saint-Porchaire, French, *c.*1550.

Plate 6. Satyr *(left)* and Maenad candelabra in ormulu on marble bases, possibly designed by Clodion and executed by Gouthière, last quarter of the 18th century.

arm-shaped sconces that appear in the sixteenth century [*figs.12 and 19*]. Human torch and candleholders (*'varlets de la chandelle'*) continued to be called on for banquets and revels: Charles VI of France is reputed to have been driven mad by the notorious *Bal des Ardens*, at which one of the torch-bearers set fire to the costumes of the dancers, and several people died. It also continued to be the practice for kings and princes to have their way lit by candle-bearers. The habit of escorting cardinals with candelabra in the French Embassy in Rome was only abolished in the 1970s. French kings, most notably Louis XIV, following their practice of rendering the French aristocracy and visiting princes subservient, were to make holding the candlestick at the royal *coucher* into the supreme honour of the ceremony. The custom lies behind our expression 'not worthy to hold a candle to' someone.

Feasting apart, the main uses for candles were for getting up before dawn and going to bed after nightfall in winter, so that candles and candle-fixtures were mostly placed on or beside chimney-ledges, where they could easily be re-lit from the embers of the fire, and on benches at the end of the bed [*pl.3*]. The upper classes – particularly their womenfolk – generally kept a night-light, or 'mortar', burning for reassurance in their room. It was customary to be lit to bed by a servant, who would then remove the light, but Margaret of Angoulême in the *Heptameron* refers to a husband and wife retaining him whilst they both read in bed! In the same century, Brantôme recorded it as unusual that Isabella of Austria kept her light in the *ruelle* beside her bed to read her book of hours by, instead of on a *buffet* at the end of her bed, like other French princesses. Medieval pictures – particularly Northern depictions of the Annunciation, whose theme justified the loving depiction of interiors, and encouraged the symbolic contrast of human lighting devices with the Light that God was sending into the world – frequently show candles in one of these locations [*fig.6*]. Sometimes the candle has no holder at all, but is held in the hand, a mere

Figure 6. The Annunciation by Rogier van der Weyden (Paris, Louvre), *c.*1435. The painting shows a brass chandelier on a pulley, and a brass sconce on a swivel over the fireplace. Similar objects are in the Victoria & Albert Museum.

rushlight or taper, or the *rat-de-cave*, the roll of waxed wick that looks like a ball of knitting and could prop itself up, though later special holders called wax-jacks were developed for it. By the fifteenth century, however, the better-off had a plentiful supply of *latten* (brass) candlesticks (forty-three are recorded at Stonor in 1472, and Margaret Paston once wrote to her husband to ask him to bring home a dozen new ones). Even for them, however, wax candles were a special luxury, reserved for honoured guests and special occasions.

Not that even tallow was cheap: the difficulty of fattening cattle meant that in times of drought or scarcity it cost four times as much as lean meat – though England benefited from plentiful mutton-fat, as a by-product of her wool and cloth trade. The best tallow candles were half mutton-fat and half beef, so as neither to melt nor break too easily. Many households (until this was forbidden when candle-taxes were imposed in 1710) made their own tallow candles, by the dipping method, and a few perhaps made wax candles, by pouring and rolling; only the best tallow candles were hard enough to be moulded, a process that can be traced as far back as 1360 in Rouen. In America in the nineteenth century it was reckoned that a slaughtered ox would yield eighty pounds of suet, furnishing tallow for three hundred candles at four to the pound. One of Thomas Tusser's *Points of Huswifery* (1570), after the caution 'Much spice is a thief, so is candle and fire', was:

> Wife make thine own candle
> Spare penny to handle.
> Provide for thy tallow, ere frost cometh in
> And make thine own candle, ere winter begin.

The wife of a small farmer or a farm-labourer made rushlights, each of which would burn about half-an-hour, out of bleached and dried pith of rushes, and – according to Gilbert White, who narrowly described the whole operation in *The Natural History of Selbourne* (1789) – 'the scummings of her bacon-pot'.

Outgoings on lighting were such a significant expense in the Middle Ages that officials of the King's household in England received both wax candles and 'paris candles' – which appear to have been special moulded candles of tallow – as part of their salary (with an additional allocation to a secretary in winter, 'when he has need of much writing'); whilst candle-ends, despite fruitless attempts to check this right up to the time of Prince Alfred, became one of the perquisites, or 'perks', of royal servants. The supply of candles led to the creation of a special post in the royal household, that of sergeant-chandler, who is significantly recorded in Edward II's household as one of the sausery, or kitchen-staff. Meanwhile, in the burgeoning city of London two new trades are recorded in the thirteenth century; those of 'cierger', or 'cirgiarius', and 'oynter', or 'unctuarius', which in the next century were formed into the guilds of wax-chandlers and tallow-chandlers respectively. One of the results of guild regulation was not merely supervision of the quality of the two kinds of candle, but also a degree of standardization, which facilitated the change from prickets (onto which any size of thick-ended dipped candle could be stuck) to sockets (for which the cylindrical end had to conform to the size of the hole). Certain candles were made with bulbous ends to stand on their own, whence their name of '*perchers*' [*fig.19*].

The Later Middle Ages and after

It has already been mentioned that brass, especially the alloy called latten (from the French *laiton*) was the usual material for candleholders in the Middle Ages, because it did not tarnish. Iron, however, was also common, whilst wood was used for the simpler cross-shaped 'candlebeam' chandeliers found in secular, as opposed to ecclesiastical, interiors. Wood was, however, both inflammable and difficult to clean. In the fifteenth and sixteenth century a notable enrichment of forms and materials took place. The vigorous export-industry of brass objects, from the Low Countries and the Meuse (now centred, until its sack in 1466, upon Dinant, whence the generic French name of *dinanderie* for its products) diversified from simple candlesticks and candelabra into wall-sconces and chandeliers – the former often on swivels at the chimneypiece, for those also profiting from the warmth and light of the fire, and the latter sometimes on a counter-weighted pulley – as several Netherlandish pictures reveal – to facilitate regulation of the candles in them [*fig.6*]. Sir John Fastolfe had a 'hangying candylstyck of laton', doubtless from Dinant, despite Richard II's and Henry IV's decrees prohibiting their importation. At the same time the

Figure 7. Persian-type candlesticks, from the Sultanate of Rum, in bronze inlaid with niello and silver, 13th century.

Figure 8. The change from pricket to socket, illustrated by three brass candlesticks: (a *centre*) conical ringed shaft, originally supported on residual tripod feet and deep grease-bowl under pricket, Flemish, 15th century; (b *left*) Persian-influenced bell-shaped foot, grease-pans under socket and half-way up shaft, English, 16th century; (c *right*) baluster shaft, still with grease-pan half-way up, Dutch, 17th century.

design of candlesticks grew more sophisticated: probably under the influence of the massive Persian inlaid bronze candlesticks [*fig.7*] which were imitated on a smaller scale by Veneto-Saracenic craftsmen in the fifteenth century [*pl.4*], bell-shaped bottoms replaced tripod feet, and sockets, with slots to facilitate removal of the candle-stubs, succeeded prickets (which were rare by the seventeenth century, save for large church candlesticks). This in turn led to the substitution of grease-pans at the foot and/or half-way up the shaft for the grease-bowls under prickets, and, as the manufacture of candles became more reliable, to their disappearance altogether, leaving only an elegant baluster stem [*fig.8*]. Humbler ones in iron, designed for practical uses like sewing or reading by, were often made with raisable spiral or ratchet sockets, in order to allow the level of the light source to be kept constant [*fig.9*]. A bizarre Late Medieval development, for which Dürer made several designs, was the trophy light – suspended antler chandeliers, with candles set on the tines, and the root carved into the torso of a woman or a monster [*fig.10*]. Gold and silver had always been employed

Figure 9 (left). A raisable spiral candle-holder (shown in the half-raised position) from Munich, steel, 17th century.

Figure 10. A drawing for a mermaid trophy-light by Albrecht Dürer (Vienna, Albertina), designed for Dürer's humanist friend Willibald Pirckheimer, dated 1513.

for special lamps and candleholders for church use, but in the Late Middle Ages and Early Renaissance royal and aristocratic love of display led to their adoption for secular use also. Whereas in 1249 the King was ordering two candlesticks of iron to be attached to the columns flanking the dais in the hall of Oxford Castle, by 1392 the Earl of Arundel was able to leave his wife 'two candlesticks of silver for supper in winter'. Henry VI is recorded as owning a pair 'of gold set with four sapphires, four rubies, four emeralds, and twenty-four pearls' – and from this time on references to exceptional candlesticks in precious metals are not infrequent. But at the same time, a great diversification into other materials occurred: candlesticks, for instance, were made in pewter in England *[pl.5]*, in maiolica in Italy *[pl.4]*, in Limoges enamel and in the rare Saint-Porchaire glazed terracotta *[pl.5]* that was virtually monopolized by the Court in France; both candlesticks and chandeliers, profiting by the added refulgence that faceting and cutting could confer, were made in crystal and glass. The will of King René of Anjou already records 'ung chandelier de verre cristallin' in 1471, but true rock-crystal chandeliers do not make their appearance till the seventeenth century, as a predominantly secular form of lighting.

Even royal chandeliers and candelabra rarely matched those of the church *[fig.11]* in intricacy and magnificence before the late seventeenth century. In quantities of lights, too, churches still outshone: no less than 400 bronze chandeliers are recorded in Antwerp Cathedral in the mid-fifteenth century, and over 300 wax candles, ranging from eight to one pound in weight, not counting those at altars and private chapels, in Tournai Cathedral in 1619. Many of these were in the twelve brass 'crowns', culminating in a triple crown in the crossing, that were a development of the Gothic chandeliers seen in fifteenth-century paintings. In its classic form, with detachable S-shaped branches emerging from a central globe and baluster, this kind of chandelier was exported from the Low Countries in such quantities that it

Figure 11. An altar-cross and candlesticks by Valerio Belli, of rock-crystal set in silver-gilt, probably made for Pope Paul III, *c.*1545.

came to be known in France as the '*lustre hollandais*'. The manufacture of this kind of chandelier and of baluster-shaped brass candlesticks was greatly aided by the application of water-power to turning in the mid-sixteenth century. In the Low Countries it was used both in houses and in churches, but in England, where its manufacture was imitated from the late seventeenth century, it was mostly restricted to church interiors. Given such a profusion of lights in churches, the Reformation, by banning them, must have had a significant effect in increasing the amount of lights used in private homes in Protestant countries like England. Gold and silver would have been melted down and forfeited to the Crown, and particularly fine bronze candlesticks doubtless accompanied the wholesale export of religious images (the Trivulzio candlestick in Milan Cathedral may well have come from England), but the hordes of lesser brass candlesticks and chandeliers without 'images' probably went into the very houses that had been built from the spoils of the dissolved monasteries.

It is, however, in the latter part of the seventeenth century, with the universal adoption by the upper classes of silver for every kind of implement and furnishing, that the real proliferation of objects connected with domestic lighting begins. Other materials were of course used, for beauty and variety, but even Wedgwood confessed the 'disagreeable truth . . . metal is the only proper candlestick material', and, of the metals, silver alone combined the properties of worth, solidity, and refulgence. Single silver candlesticks diversified

Figure 12. A silver-gilt sconce, with arm-shaped support, I.L. (London), 1684.

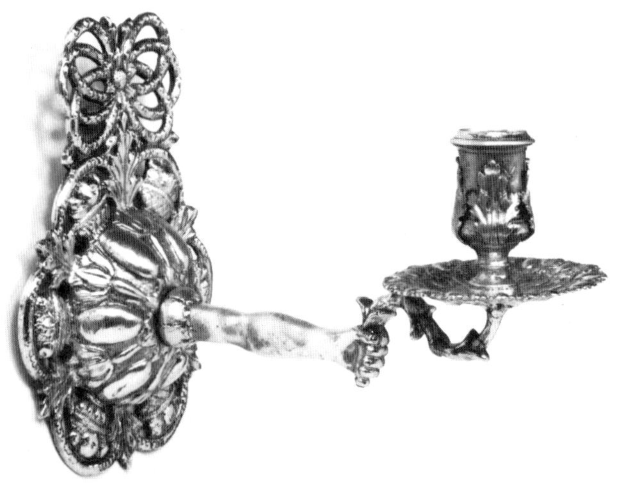

Figure 13.
A gilt-wood girandole, supplied by Thomas Chippendale to the Duke of Portland in 1765–6.

from the baluster shape normal in brass into the clustered tubular shafts popular in the mid-seventeenth century, and thence into the classic columnar form; silver candlesticks with two or more branches – commonly referred to as 'candelabra' – made their appearance *[pl.6]* (though in France under Louis XIV only the King was allowed a candlestick with more than one socket); and silver sconces *[fig.12]* (though the young Pepys was still content with pewter ones for his stairs and entrance hall in 1661–2) attached to the wall developed out of the practice of nailing a reflecting plate of brass or copper behind a candle to increase its refulgence (Evelyn remarked on 'a neate invention for reflecting lights by lining divers sconces with thin shining plates of gilded copper' as a novelty in the hôtel de Liancourt in Paris in 1644). Light-brackets – called 'branches' and, later, 'girandoles' *[fig.13]* – were attached to the sides of mirrors to produce the same effect (the use of mirrors for furnishing, above fire-places and on walls, and in special mirror-cabinets, was itself partly governed by the desire to multiply light, and facilitated by the production of larger sheets of plate-glass). Special pieces of furniture were developed for this new profusion of lights – the *guéridon* (which, when it first appeared in the mid-seventeenth century, was a stand supported by a figure of the legendary blackamoor servant from whom it took its name, though it later referred to any kind of stand, and then to an occasional table) and the *torchère* for the candelabra for social occasions, and tripod-tables for occasional candlesticks. Great magnates commissioned silver chandeliers (in England no one below a duke seems to have owned one), but rock-crystal chandeliers aroused the greatest admiration. Anne of Austria and Mazarin appear to have been the first to have had them: how precious they were is shown by Mazarin's being hung from 'silver chains wrapped in crimson silk, with gold and silver tassels at the ends'. Only slightly less costly imitations were made in crystal glass from Milan, Bohemia and France itself. In the eighteenth century *ormulu* [*pl.6*] became a favourite material in France for richer chandeliers, girandoles, and the like, partly because of Louis XIV's sumptuary laws (after melting down his own silver in 1689) prohibiting them in precious metals, in order to keep these in circulation as coinage to pay for his wars. In England intricately carved and gilded wood was popular for chandeliers and girandoles [*fig.13*], whilst from the 1760s Sheffield plate was increasingly used for industrially-made candlesticks and candelabra. Glass chandeliers reached the greatest heights of expense and virtuosity. There are occasional mentions of crystal chandeliers in England in the seventeenth century (three that Celia Fiennes admired at Hampton Court are still there), but in 1714 the first advertisements for 'Glass Schandeliers' were put out by John Gumley, and in the course of the eighteenth century English chandeliers, made of the brilliant lead glass invented by Ravenscroft, whose prestige came to be such that the French referred to 'crystal d'Angleterre' as a paragon of excellence, reached unparalleled heights of lavishness and design [*fig.14*]. In the late eighteenth century a special form of candlestick with pendant

Figure 14.
A crystal-glass chandelier from Wroxton Abbey, *c.*1815.

Figure 15.
A silver-gilt chamber-candlestick and extinguisher by John Schofield 1791–2, with fitted snuffers by William Bayley, 1792–3.

drops of glass, called a lustre, emerged as a habitual element of the *garniture de cheminée*; by the nineteenth century it had become an ornament in its own right, without candle-sockets.

On a humbler lever, special chamber-candlesticks, often in silver, were produced for lighting one's own way to bed, complete with a handle, conical extinguishers and a circular tray on which to put the snuffers and the implements for relighting them [*fig.15*].

Snuffing, Douting, and Lighting

In the Middle Ages servants were expected to extinguish candles and keep them from smoking, apparently with their fingers, but as the desire for privacy grew, their employers required instruments with which to snuff them themselves. Ever since Byron wrote of Keats:

> 'Tis strange the mind, that very fiery particle,
> Should let itself be snuff'd out by an article

the verbal confusion between 'snuffing' (the act of trimming a wick) and 'extinguishing' has been ineradicable. The flame of an oil-lamp was regulated by manipulating the wick to protrude more or less by means of tongs and tweezers [*pl.7*]; wax candles could be made with the thickness of wick and wax so matched that both were consumed at the same rate, with the wick neither burning so vigorously as to starve itself of fuel and smoke, nor so feebly as to melt more than it consumed and be doused, or to cause waste by the surplus running down the sides. Tallow melted much more easily, which left the wick high and dry and smoking, till it toppled over and its flame caused a breach in the side down which the tallow ran away, guttering, leaving the wick starved and smoking once more. A tallow candle – according to an admittedly partial French eighteenth-century advertisement – needed snuffing eight to ten times an hour; unsnuffed, according to tests made by Count Rumford, it forfeited almost two-thirds of its brilliance within eleven minutes.

Figure 14.
A crystal-glass chandelier from Wroxton Abbey, *c.*1815.

Figure 15.
A silver-gilt chamber-candlestick and extinguisher by John Schofield 1791–2, with fitted snuffers by William Bayley, 1792–3.

drops of glass, called a lustre, emerged as a habitual element of the *garniture de cheminée*; by the nineteenth century it had become an ornament in its own right, without candle-sockets.

On a humbler lever, special chamber-candlesticks, often in silver, were produced for lighting one's own way to bed, complete with a handle, conical extinguishers and a circular tray on which to put the snuffers and the implements for relighting them [*fig.15*].

Snuffing, Douting, and Lighting

In the Middle Ages servants were expected to extinguish candles and keep them from smoking, apparently with their fingers, but as the desire for privacy grew, their employers required instruments with which to snuff them themselves. Ever since Byron wrote of Keats:

> 'Tis strange the mind, that very fiery particle,
> Should let itself be snuff'd out by an article

the verbal confusion between 'snuffing' (the act of trimming a wick) and 'extinguishing' has been ineradicable. The flame of an oil-lamp was regulated by manipulating the wick to protrude more or less by means of tongs and tweezers [*pl.7*]; wax candles could be made with the thickness of wick and wax so matched that both were consumed at the same rate, with the wick neither burning so vigorously as to starve itself of fuel and smoke, nor so feebly as to melt more than it consumed and be doused, or to cause waste by the surplus running down the sides. Tallow melted much more easily, which left the wick high and dry and smoking, till it toppled over and its flame caused a breach in the side down which the tallow ran away, guttering, leaving the wick starved and smoking once more. A tallow candle – according to an admittedly partial French eighteenth-century advertisement – needed snuffing eight to ten times an hour; unsnuffed, according to tests made by Count Rumford, it forfeited almost two-thirds of its brilliance within eleven minutes.

The consequences of inattention are amusingly illustrated in Maurice Quentin De La Tour's pastel portrait of the Abbé Huber so absorbed in reading, that he has let one of the two candles gutter away to its own extinction [*fig.16*]. Not for nothing did Goethe proclaim

> . . . 'Twould be a service to mankind enough
> To invent a light that burnt without the need to snuff

– a wish finally realized with the employment of Cambacère's 'snuffless wick' in 1820.

Until that time special scissors, called snuffers, were needed to keep wicks in trim. It would seem that there was no special instrument for snuffing until pivoted scissors were invented in the Late Middle Ages, and shortly afterwards adapted to the purpose of snuffing by the addition of a box to the blades to hold the cut-off bits of wick. The earliest

Figure 16 (left). An etching after Maurice-Quentin de la Tour's pastel portrait of *The abbé Huber reading* (Musée La Tour, Saint-Quentin), from Adolphe La Lalauze, *L'oeuvre de Maurice-Quentin de La Tour*, Saint-Quentin, 1882.

Figure 17. A silver-gilt snuffer made for the court of Edward VI (1547–53), inscribed: 'GOD SAVE THE KYNGE EDWARDE WITHE ALL HIS NOBLE COUNCEL'.

snuffers, e.g. the silver pair made for Cardinal Bainbridge in Rome around 1510 (British Museum), or Edward VI's (1546–52) exquisitely decorated silver-gilt pair in the Victoria and Albert Museum *[fig.17]*, have a heart-shaped box at the end of the blades; later snuffers have a rectangular box in the middle. For putting candles out, scissors with a pair of flat discs at the end, called douters, were an alternative to conical extinguishers. These instruments were not always conveniently to hand, so that fingers still had to serve both for snuffing – as when Henry IV made his son Louis XIII compete with a favourite to be his *mignon* by being the first to snuff a smoking candle – and for extinguishing; those trying to avoid burning their fingers doing the latter devised a host of bizarre or perilous alternatives – from throwing their shirt over (strongly discouraged by the fourteenth-century *Ménagier de Paris*), to the litany of methods detailed by Swift in his *Directions to Servants*, including the ironical advice: 'You may run the Candle end against the wainscot . . . you may whirl it round in your Hand till it goes out: when you go to Bed, after you have made Water, you may dip your Candle End into the Chamber Pot . . .'

It was easy when snuffing, particularly without the proper instruments, to put the candle out inadvertently (which is how 'snuffing out' came to mean 'extinguishing') and, as Boswell records when in London, it was not always so easy to relight it:

I determined to sit up all night . . . about two o'clock in the morning I inadvertently snuffed out my candle, and as my fire before that was long before black and cold, I was in a great dilemma how to proceed. Downstairs did I softly and silently step to the kitchen. But, alas, there was as little fire there as upon the icy mountains of Greenland. With a tinder box is a light struck every morning to kindle the fire, which is put out at night. But this tinder box I could not see, nor knew where to find. I was now filled with gloomy ideas of the terrors of the night. I was also apprehensive that my landlord, who always keeps a pair of loaded pistols by him, might fire at me as a thief. I went up to my room, sat quietly until I heard the watchman calling 'past three o'clock'. I then called him to knock at the door of the house where I lodged. He did so, and I opened to him and got my candle relumed without danger. Thus was I relieved . . .

As Boswell makes clear, the commonest method of getting a light was to procure it from elsewhere – from another candle, a fire, or even a passer-by. Since this was not always possible, the universal stand-by was the tinder-box. This consisted of a container holding the essential implements and ingredients for procuring a flame: a fire-steel and a flint to produce a spark; tinder (usually of charred rag prepared by the servants, but the best was *amadou* – the French word for tinder, derived from *amadouer*, to wheedle or coax – or German tinder, made of the dried tree-fungus *Boletus igniarius* steeped in saltpetre) to take this spark and smoulder; and often a sulphur-match to make these embers break more easily into flame. A late work of fantasy, *The Tinder Box* (1832), reckoned that 'there are very few house-men, or house-maids, who can succeed in "striking a light" in less than three minutes' – and this was doubtless the case in the dark. The rich could use the tinder-pistol, working on the same principle as the wheel-lock pistol in an ingenious adaptation first made in the seventeenth century. Not until the 1780s was phosphorous used to make the first chemically combustible matches, ignited by dipping; and only in 1826–7 were these perilous devices succeeded by the invention (of John Walker from Stockton-on-Tees) of the friction match, which it was left to others to patent, improve and exploit as 'lucifers' and 'congreves' (after the rocket inventor); these were the ancestors of our own safety-matches (first made in Sweden in 1855). The curious thing about fire-making implements is how rarely, compared with candle-holders and snuffers, they or their containers were the objects of craftsmanship and display. This was partly because they required rough handling and got dirty, and partly because getting a light was a once-and-for-all operation, so not for show; but it was also because they were handled by the menials, who made fires and procured lights before the leisured members of a household were stirring, and brought lights in the evening.

For the great enrichment and proliferation of light-holding devices that we have been examining were not the result of utilitarian pressures for better lighting, but of the evolution of a way of life whose chief objects were entertainment and display; to a great extent as ends in themselves, but also as instruments of social differentiation. By proliferating the number of lights required on social occasions – in private, in the bosom of the family, the single candle still usually sufficed to read, sew or go to bed by [*fig.18*]; by pushing the hours of their day and amusements ever further into the night, requiring the extended use of lighting; and finally, by insisting on the more delicate and reliable, but costly, candle of wax rather than tallow, the leisured classes established a mode of life that only they could afford, and that added to the increasing number of distinctions setting them off from those whose life was governed by work and its rhythms.

Meal-times and bed-times

'Early to bed and early to rise' and 'Rise with the lark, to bed with the lamb' were counsels not so much of moral perfection as of economy for the poor (or rather, as with the temperance movement, the two were confused). Their day was so ordered as to minimize the use of artificial light, or to dispense with it entirely. Even Anthony Fitzherbert's *Book of Husbandry* (1523) addressed to the literate farmer, admonishes:

> One thinge I wyl advise the to remembre and specially in wynter-tyme, whan thou syttest by the fyre, and hast supped, to consyder in thy mynde whether the workes, that thou, thy wyfe, and thy servantes shall do, be more advantage to the than the fyre, and candell-lyghte, than syt styll: and if it be not, than go to thy bedde and slepe, and be uppe betyme, and breake thy fast before day, that thou mayste be all the short wynters day about thy busyness.

This advice was reflected in Thomas Tusser's *Points of Husbandry* (1577):

> In winter at nine, and in summer at ten
> To bed after supper, both maidens and men.
> In winter at five a'clock, servant arise
> In summer at four is very good guise

Nor, in the short winter days, did rising in the dark and supping after night-fall necessarily involve the use of light, or a medieval writer would not have felt obliged to say 'It is a shame to suppe in darknes, and perillous also for flyes and other filth'.

Originally, when masters and men dined together in the great hall and slept in the same chamber, their hours – save for the requirement that servants rose first and turned in last – were necessarily the same, and will not have differed much from those of the rest of the common people. Save for the occasion of some special feast, no one delayed his evening repast much beyond dusk, whilst the general existence of a curfew more or less ruled out social activity outside his own family circle for the ordinary man. It seems to have been around 1500 that the upper classes of Northern Europe began to change their hours of sleeping and eating, possibly under the influence of an earlier revolution in manners in Italy (certainly Montaigne was to find supper in Italy agreeably late). Louis XII of France (1498–1515) was reputed to have hastened his death by dining at noon instead of at eight, and sitting up till midnight instead of six. The new hours, with the main meal of the day – dinner – at eleven, or at latest, at noon, continued until the reign of Louis XIV, when life took another lurch into the night. By 1710, according to Liselotte, though dinner was only a little later, at 1 p.m., supper was not until 10 p.m. or quarter to eleven; and during the course of the century the ways of the French court were steadily aped all over Europe and exaggerated yet further, as unpunctuality became a mark of fashion. In 1777 a testy Horace Walpole observed:

Silly dissipation rather increases, and without an object. The present folly is late hours. Everybody tries to be particular by being too late; and, as everybody tries it, nobody is so. It is the fashion now to go to Ranelagh two hours after it is over . . . Lord Derby's cook lately gave him warning. The man owned he liked the place, but said he should be killed by dressing suppers at three in the morning.

The social arrogance of such behaviour was pinpointed by Sébastien Mercier: 'Life by candlelight is really a manifestation of opulence . . . all the rich have taken against the sun . . . Certain women in Paris only get up towards evening and sleep when dawn appears; a fashionable women generally adopts this habit, and is referred to as a "lamp" '. The later, but less exaggerated, hours of even the ordinary middle class are typified by Fanny Burney's account of her family's life at King's Lynn in 1768: 'We breakfast always at ten, and rise as much before as we please, – we dine precisely at two, drink tea about six – and sup exactly at nine.' Once she became Assistant Keeper of the Robes to the Queen in 1785, however, her dinner was at five and supper not till eleven.

By the mid-nineteenth century dinner had moved into the early evening, so as to displace supper and to call into existence cold luncheon in the middle of the day, with tea between; upper-class hours were thus set more or less into the present mould, but for the transformation of cold 'luncheon' into the more important hot 'lunch', and the re-substitution of the less formal supper for dinner, as servants dropped away. With artificial light now within reach of all, and with no leisured class exempt from the (at any rate felt) necessity to work, differences in hours kept have been largely eroded; so that now the whole nation has to have Summer Time to save artificial light. The idea of Daylight Saving (first introduced in 1916) was indeed suggested to its promoter William Willett on his return from an early summer morning's canter, by the sight of so many large houses with their blinds drawn down against the daylight.

4 Wax versus Tallow

Hand in hand with the assertion of prodigality in the shape of late hours went conspicuous consumption and social distinction in the number and quality of lights burned. In the ancient world, as Martial's Epigrams reveal, the chief social distinction was between the rich who used lamps *(lucernae)* and the poor who had to be content with some kind of candle or rush-light (a *candela*, or a *cicindela*); he contrasts the profligate contemporary use of lamps by the Romans, with the frugal adherence to candles of their ancestors. These candles were generally of tallow; the *cereus*, or wax torch, was generally rather more special, being brought, for instance, as a gift from clients to their patrons at Saturnalia. In the Middle Ages, when Northern Europe went over to candles for domestic lighting, leaving lamps to the Mediterranean countries where olive oil was plentiful *[pl.7]* (transalpine artists like Trophime Bigot and Honthorst who went to Italy in the early seventeenth century were to delight in their contrasted effects), or to outhouses and to the poorest classes, who burnt inferior substances such as tripe-oil in devices like the Scots crusie, wax was such a precious commodity that only the Church insisted upon its exclusive use. A pound of beeswax cost as much as a labourer's day's wage in the Middle Ages, and was always three to four times as expensive as tallow. Even the King used wax exclusively only in his bed-chamber – including two wax 'mortars' to burn throughout the night. This preference for wax candles in the bed-chamber did not remain confined to royalty: the last of the entertaining dialogues in Peter Erondell's text-book *The French Garden* (1605), called 'which treateth of going to bed', has the housekeeper say: 'Take out of the way this Pewter-Candlestick which is so foule, make readie the Silver

Candlestickes with the waxe candles, for you know that she cannot endure the smell of tallow, because it doth most often stinke.' In other rooms, even royalty sometimes yielded to penury or parsimony: the sovereign Counts of Foix dined by tallow candles held by servants in the fifteenth century; the Emperor Leopold I of Austria was almost poisoned by the fumes of the arsenic used to bleach his tallow candles in the seventeenth century; and in the eighteenth century the Margravine of Bayreuth acidly recorded her father's meanness in using tallow candles, which blackened the solid silver furnishings of the Stadtschloss of Berlin – not to mention people's clothes and faces. That tallow candles were compatible with making a show in less exalted circles is brought home by Stendhal in his *Vie de Henry Brulard*: 'My aristocratic family would have felt ashamed if the candelabra had not been of silver. It is true that the candles they bore were not of the noble wax – it was customary then to use tallow. But – these tallow candles, they came in specially packed chests from near Briançon, and they had to be made of goat fat.' Even in England in the early nineteenth century, where foreigners were generally astonished by the generally high living-standards and the luxurious provision for travellers, Prince Pückler-Muskau noted that even in the best inns tallow, and not wax, candles were used and that if you insisted on wax the bill was doubled (though in 1834 the Quaker Caroline Fox records virtuously substituting 'moulds', moulded tallow candles, for the wax ones automatically supplied at an inn in Salisbury). In private life in England and France, however – or at least in their respective capitals – wax was virtually *de rigueur* for the upper classes by the mid-eighteenth century (and was taxed more heavily accordingly). In the Middle Ages it was already the custom to accord special treatment to guests by giving them wax candles, whilst Philip the Fair forbade the use of wax torches to bourgeois and clerks. There were, of course, practical advantages of wax over tallow, which were the ones that persuaded Pepys to try the change from one to the other in 1664. As

Thomas Fuller remarked, when listing it under the 'commodities' of Hampshire (Doncaster was another great centre of production), 'Wax is good by day and night, when it affordeth light, for sight the clearest, for smell the sweetest, for touch the cleanliest': and what was more, wax candles rarely needed snuffing, so that their use became almost inevitable in theatres and ballrooms *[pl.8]*, where the number of lights and the activities made constant attention almost impossible. But, as Shakespeare's simile implies:

> Base and unlustrous as the smoky light
> That's fed with stinking tallow

the disadvantages of tallow – dirt, smell, lack of brightness – were of the very kind to condemn it in social terms as well. In France a separate word, '*bougie*', to distinguish the wax candle had already been coined in the fifteenth century; it was a distortion of *Bujiyah*, the name of the port in North Africa from which the best wax came (as had the *Punica cera* of Roman times). The word '*chandelle*', which was originally applied to any form of candle, gradually came to mean only one of tallow. The distinction between the two words and the two substances was not however absolute even in the eighteenth century. When Louis XIV died in 1715, Dangeau was still able to write 'The King died this morning . . . he gave up his spirit without any struggle, like a *chandelle* going out.' It was only his editor later in the century who thought it necessary to substitute the word *bougie*, lest it should be thought that the Grand Monarque, like Falstaff and other mortals, was: 'A wassail candle, my lord – all tallow.' The delicate social problem involved in whether to accord someone else wax or tallow was nicely revealed when a new tutor came to Leinster House in the mid-eighteenth century. When the Groom of the Chamber asked which he should be given, he was told 'Oh, moulds (i.e. tallow) will do, till we see a little.' He was doubtless soon advanced to wax, for in the fullness of time he was to marry the widowed Duchess. Horace Walpole, visiting Lady Mary Wortley Montagu on her return to England in 1762, was shocked to find her 'in a little miserable bedchamber of a ready-furnished house with two *tallow* candles and a bureau covered with pots and pans'.

The increase in the consumption of wax that the extension of its use downwards in society gave rise to is recorded with concern by the *Encyclopédie* in 1772: 'It has become such a great necessity in several arts and crafts and in domestic life, that its consumption is scarcely credible – particularly now that its use is no longer exclusively reserved to churches and royal palaces, and that everyone now uses *bougies*, Europe can no longer furnish its own needs . . .' Crisis – or an ignominious reversion to tallow – was only avoided by technical advances in the next decade, which made oil-burning lamps once again more attractive.

Numbers of lights

The decision to use wax or tallow was governed for those who were held back from the expense of constant wax candles by whether or not company was being formally received; this also dictated the number of candles. As has been said, the individual, or even the family alone, tended to make do with a single candle, or at most two. This can be very well seen in a rare set of drawings of family life from the first decade of the nineteenth century by an amateur artist and friend of Constable, John Harden of Brathay Hall, Westmorland, now in the Abbot Hall Gallery, Kendal [*fig.18*]. Though a country gentleman with an adequate income, Harden shows his family circle of four all reading and sewing together in the evening by the light of the fire and a single candle in one drawing, and by two candles only in others. At a slightly humbler level of society, the poet Crabbe says of the household of his great-uncle, a farmer worth £800 p.a., that except on 'great and solemn occasions . . . the family and their visitors lived entirely in the old-fashioned kitchen along with the servants . . . Mrs Torell sat at a small table on which, in the evening, stood one small candle in an iron candlestick, plying her needle by the feeble glimmer, surrounded by her maids,

Figure 18. Family occupations by candlelight (1804, Abbot Hall Gallery, Kendal), by John Harden (1772–1847).

all busy at the same employment . . .' Even Lady Leicester at Holkham was reported as working at her tent-stitch frame every night by the light of a single candle in 1774. It was on the other hand noted as a peculiarity of Napoleon's (for instance, by Mademoiselle George, the actress, who affected bashfulness at so much light) that he liked to live in an imperial blaze of light '*comme un jour de bal*'. Meagre everyday lighting in a large house did not reach to the corners of a room, or Fanny Price on a sofa at the other end of the drawing-room would not have been invisible to her cousins when they came back to *Mansfield Park* after a late ride. Later in the book, when she visited her own poor family (who nonetheless had servants) at Portsmouth, Fanny Price found that the solitary candle in the house was hogged by her father reading the newspaper. Sometimes no light at all was forthcoming; when Pitt increased the candle taxes in 1784, it was on the calculation that poor families used no more than ten pounds by weight per year.

Lack of light was a particular affliction of poor scholars

who needed to study far into the night, 'burning the midnight oil' (this last curiously a phrase coined by Francis Quarles in his *Emblems* of 1635, when oil was hardly in use in England). Henri III's tutor Jacques Amyot claimed that he had been forced to study by the light of the votive lamp kept perpetually burning before the image outside his lodgings. William Oughtred, the inventor of the slide-rule and of 'x' as the multiplication sign, was less lucky: John Aubrey (who admits that he did not rise till midday) says that 'his wife was a penurious woman and would not allow him to burne candle after supper, by which means many a good notion is lost'. In the case of the Brontë family it was the old servant Tabby who was opposed to a candle being lit, so that the girls used to pace up and down in the firelight talking, planning, and discussing the plots of their novels.

So, too, parts of a house where no one remained, like stairs and corridors, remained simply unlit, even in the Tuileries under Napoleon. Receiving a visit always required additional light, as a gesture rather than from necessity. Walpole told Mann the story:

> I remember an old superstititious parson by Cambridge who met his daughter one night going up to bed with a farthing candle in her hand; he asked her where she was going? She replied, to bed. Well, says he, and you design to say your prayers first, I'll warrant you! Yes, Sir. Yes, Sir, and you are going to talk to God with only a farthing candle? If any foolish visitor or gossiping madam was to come to you, you would light up two tallow candles – pray go light up two to say your prayers by.

The notoriously parsimonious Nollekens claimed 'that a pair of [tallow] moulds, by being well nursed, and put out when company went away' once lasted him a whole year.

The real occasion for multiplying lighting was when entertaining. Montesquieu carefully charted the distinctions in frugal Florence: 'Life in Florence is very economical . . . in houses when there is no gaming, one is lit by one little lamp; when a few people gather, by a large lamp; and when company enters, by lighting all three spouts of this – for it has three, and rests on a sort of lamp-stand' [*pl.7*]. In England, where, as Swift observed in the early years of the eighteenth century: 'You cannot but observe of late years the great extravagance among the gentry upon the article of candles', things were done more magnificently, as by the *Spectator*'s caricature, Lady Enville, who 'makes an Illumination once a week with Wax-Candles in one of the largest Rooms, in order, as she phrases it, to see Company'. The number of lights depended above all upon the wealth of the host, but also upon the standing of his guests and the impression he was out to make [*fig.19*]. When Sir Robert Walpole entertained the Duke of Lorraine in the uncompleted Houghton in 1731 'They dined in the hall, which was lighted by 130 wax candles, and the saloon with 50.' These were the occasions

Figure 19. Candles for an occasion of public rejoicing: Charles II banqueting in the Hague in 1660. Note the perchers placed as illuminations in the windows, the early crystal chandeliers, the cluster-shaped candlesticks, and the (empty) arm-sconces.

Figure 20. A Venetian carved-wood hanging-lantern, from the Palazzo Gradenigo, *c.*1570. The lantern's glass panes have been lost.

for the use of hanging-lanterns [*fig.20*] and glittering chandeliers; when the one in the great hall at Bulstrode was lit for the visit of George III in 1779, it was the first time for twenty years. As for the host of other sconces, girandoles, and candelabra dotted about on tripod tables and torchères, it was perfectly common to hire them for the occasion, even at Versailles [*pl.8*]. The heat generated by such a prodigious amount of people and burning candles can easily be imagined; when the dancers complained of the heat at an ostentatious ball given by Sir Laurence Dundas in 1769, his wife nonchalantly ordered the servants to break a pane of the fine crown glass in each window to let the air in; in Westminster Hall at the Coronation of George IV the heat was such that the candles in the twenty-eight glittering chandeliers wilted and dropped hot wax onto the extravagant finery of the guests. A further drawback was that, as Sir Frederick French

Plate 7. A triple-spouted raisable lamp, with attached tweezers, snuffers, douters, and prodder, Urbino maiolica, by G. A. M. Roletti, 1773.

Plate 8. Giuseppe Grisoni (1669–1769), *A masquerade at the King's Opera House in the Haymarket*, c.1720. It is reported that at one such ball 500 wax lights were used.

Plate 9. Piccadilly Circus by night: neon advertisements, street lighting, and car headlamps.

Plate 10. A princess celebrating Diwali. Mughal miniature by Hūnhar, *c.*1760. Note the adoption of Western candelabra and lanterns.

Plate 11. The Duke of Richmond's fireworks, 15 May 1749, organized by Charles Frederick, 'Comptroller of His Majesty's Laboratory at Woolwich'.

Plate 12. Design for a 'period' panelled drawing-room, with typical converted vase table-lamps on a side-table and torchère, and sconces over the fire-place. Anonymous, *c.*1930, for William Henry Haynes & Co., 'Upholsterers & Antique Dealers'.

calculated, 'one wax candle consumes as much oxygen as two men.'

Fire danger

The use of an open flame for lighting was, of course, conducive to things catching fire, not only houses but people. Pepys records setting his periwig alight when using a candle to seal a letter in 1668. Girodet (like the short-sighted man in Hogarth's print) set fire to his hat when reading by candlelight one evening. Most hazardous of all was reading in bed: it was the cause of as many fires as smoking in bed nowadays. A medieval sacristan of Durham set fire to himself doing this; and as late as 1889 Price's were exhibiting 'Timed Safety Bed Candles: prepared to burn 30 minutes only with a view to the prevention of the dangerous habit of reading in bed'. As to buildings, statistics reveal that in 1835 110 out of 471 London fires were caused by candles, and in 1848 237 fires out of 767 – but another 65 were caused by the erratic gas-lighting that had by then begun to appear.

Lesser hazards of candles and lamps included dripping wax and oil. Swift's Directions to Servants included the ironic advice to save themselves trouble by discouraging the use of too many candles – for instance, by placing a candle in a sconce 'so as to make it lean, in such a manner, that the grease may drop all upon the floor, if some lady's head-dress or gentleman's periwig be not ready to intercept it'. Mercier noted dripping wax from the torches of noblemen's coachmen as one of the many traffic hazards in Paris, whilst Hogarth's *Arrest of the Rake* shows the dangers to passers-by from the refilling of the street-lamps. Oil-lamps offended much worse than candles in this respect, for capillary action drew excess oil to drip over everything, unless – as with the crusie – there was a second receptacle to catch the excess. Greuze was known as 'The Bucket' as a student at the Academy, because, being poor and humble, he was relegated to the back of the class, where he caught the drips from the lamp illuminating the model.

5 Public Lighting

Night-life: lights for show and entertainment

Entertainment was the great generator of lighting, not only in the home, but in public as well. As civil commotions and tumults eased with the reinforcement of royal power at the end of the Middle Ages, curfews were lifted, new forms of diversion grew up, and began to extend into the evening, and – in order both to make it easier to control the participants in this increase in nocturnal activity, and to make it safer for them – street-lighting improved. Street-lighting originated in the conversion of the lights put out by householders to celebrate church festivals or state events in the Middle Ages into lanterns hung out compulsorily on the moonless nights of winter. De la Reynie, the first *Lieutenant de Police* in Paris, was the earliest to organize this centrally, in 1666; whilst the City of Westminster was the first to undertake street-lighting every night of the year, in 1735–6. Deficiencies in street-lighting still made it necessary to carry a portable lantern or have one's way lit by servants or a link-man or -boy with torches (from the Latin *torquere*, because these were made of strands of some substance like tow, impregnated with fat or wax, twisted together). The reflectors used to concentrate the beam of street-lamps in the seventeenth century were soon adapted to carriage-lanterns, but these did not project light far enough forward to travel at all fast across country, and even someone like Parson Woodforde could not afford his own. For horses there was no solution: the Lunar Society (1765–91) derived its name from the fact that its meetings were held at the full moon, so that its members could return home after dark; whilst Dr Bovary had to wait till moonrise before he could visit a farmer with a broken leg. Only with the arrival of street lighting by gas (first tried out in Pall Mall in 1807), and then electricity (with Edison's creation of the first public electric supply station by the Holborn Viaduct in 1882) could towns be lit brightly enough to ensure safe movement for the footgoer and, when it appeared, the motor-car (off whose dynamo headlights could be run to allow fast travel in the open) *[pl.9]*.

These improvements in street-lighting went hand in hand with an increase in not only private but also public gatherings, and in entertainment after dark. Taverns were the first places to encourage casual gregariousness; originally restricted by law (as in Scotland on Sundays still!) to supplying *bona fide* travellers, by 1600 – when the first governmental concern about public drunkenness was expressed in France (already tempered by caution over the forfeiture of revenue that would ensue from doing anything radical about it) – they had become places of casual resort for drinking. The seventeenth century added tea-houses, coffee-houses and eating-houses, as well as multiplying places of indoor entertainment for the public, instead of just the court-like cockpits, theatres, opera-houses, and the first public concerts. The eighteenth century saw the development of coffee-houses into clubs, the establishment of special clubs for gaming, and the building in virtually every English provincial town of assembly-rooms and ball-rooms *[pl.8]* – where previously churches and cathedrals had been the only indoor places for mingling – and in London the creation of such places of entertainment as Vauxhall, Ranelagh and the Pantheon, where the brilliant lighting was one of the attractions: both Evelina and Lydia Melford found Ranelagh 'like the enchanted palace of a genie'. The change in role that this involved for the candle is pungently illustrated by Thomas Dekker in *The Seven Deadly Sinnes of London* (1606), who, after listing its worthy former uses, now rounds on it with: 'And art thou now become a companion for Drunkards, for leachers, and for prodigalles? Art thou turned Reprobate?' Much later, after the arrival of gas but before cheap lighting was generally

Figure 21. A Polish brass Hannukah light, 18th century.

available, Dickens remarked in his *Sketches by Boz* (1835) that the 'perfectly dazzling . . . profusion of gas-lights in richly-gilt burners' was one of the great lures of the new 'gin-palaces' for those living in the drab and dirty housing of the poor.

Lights for rejoicing

The use of lights in quantity in the past was, as has been said above, always for demonstrative rather than practical reasons. The Egyptians and Romans had festivals of lights and the Jews and Indians still have their Hanukkah *[fig.21]* and Diwali *[pl.10]* celebrations. The Romans also used to put lights out of doors to celebrate occasions of private and public rejoicing – and in the later Middle Ages lights were ordered to be put out to mark the arrival of a prince in the town. This was partly for public order (and hence, as we have seen, one of the sources of street-lighting), but also to express joy. The practice survived the Middle Ages *[fig.19]*: Evelyn records coloured paper lanterns being put at the windows in Italy; Anthony à Wood mentions candles at the windows of Merton when James II succeeded to the throne – three years later seven candles would be set in Whig windows to celebrate the acquittal of the non-juring bishops; in 1782 Walpole had his windows broken by the mob for not having lights in them to celebrate Rodney's victory in the Battle of the Saints, commenting cynically that 'glaziers and tallow-candlers always treat sashes as public enemies' (in 1859 'the terrorism of glaziers and gasfitters' was held to blame); and to this day flambeaux are lit outside the clubs of Pall Mall and elsewhere to celebrate royal marriages.

Quantities of outdoor lights contributed to the gaiety of both private and public gatherings. At Bussy-Rabutin's party for his cousin Mme de Sévigné, the gardens of the temple were illuminated by a hundred crystal chandeliers hanging from the trees; the Duke of Brunswick put out 5000 saucer-lamps at an open-air party in Herrenhausen in 1741. All the garden entertainments – Vauxhall, Cremorne, and Ranelagh – were noted for their multitude of plain and coloured glass lights amongst the trees. The lineal descendant of these displays are the illuminations of Blackpool and other coastal resorts.

Bonfires were an additional way of rejoicing with light and flame from early times, as were, with the invention of fire-arms, salvoes of artillery. It was, therefore, natural that fireworks, which combined all three elements of rejoicing – brilliant (and coloured) light, sparks, and explosive sound – should have been taken up so enthusiastically in Europe after their introduction from the East. In India lavish displays continue to be mounted at the light-festivals of Diwali *[pl.10]* and Dasserah. In Europe, the German tradition of letting off great displays of squibs and rockets into the sky has displaced the Italian tradition of architectural set-pieces that prevailed until the mid-nineteenth century *[pl.11]*: it is tempting to suggest that this is connected with habituation to sheer quantity and brightness of light (thus also accounting for the disappearance of prodigal displays of lighting at funerals – see Nicholas Penny's *Mourning* in this series), so that only the dramatic and illusionistic use of light retains its capacity to delight and astonish us – as it does in the theatre, *son-et-lumière*, and laser light-shows.

6 Technical Advances in Lighting

For several millennia, whatever the increases in the number of lights used and the elaboration of their supports, there were virtually no significant technical advances in lighting. Lighting was inevitably dim, and this was one reason for the medieval guilds' prohibition of the making or the sale of goods by artificial light. The solution to one problem, that of ensuring that the amount of fuel was kept proportionate to the flame, to avoid smoke and repeated snuffing, was found (for the lamp) in the papers of the sixteenth-century thinker Cardan in the seventeenth century, and exploited as the 'philosophical lamp', which had a counterpoise to keep the oil level steady (in the version that Evelyn saw demonstrated to the Royal Society in 1667 the declension of the counterpoise also recorded the time). For candles the answer was the plaited wick, which inclined the edge of the wick into the hot outer edge of the flame, where it was burned away. Sébastien Mercier ecstatically announced the invention of plaited wicks for oil-lamps by Léger to his fellow men-of-letters, '*veilleurs déterminés*', in 1783, but it was not until 1820 that Cambacères applied the plaited wick to the candle, and its operation was not ideal until the introduction of the glycerine-rich stearine candle in the 1830s, which resulted from the publication of Chevreul's researches into the isolation of fatty acids in 1823.

The eighteenth century witnessed a mass of new designs for lamps, but the first major advance to produce an increase in the intensity of the light-source was the Argand lamp, which the Swiss distiller, Ami Argand, perfected between 1780 and 1784. He made two innovations: the wick was hollow and enclosed between two tubes, so as to introduce oxygen to burn the oil vapour on both the inside and the outside of the flame, which reputedly made it as bright as ten to twelve tallow candles; and a glass chimney was placed above the lamp to increase the draught and intensify the brightness yet further (as Leonardo, using metal chimneys, had observed long before) – supposedly to that of twenty candles. Unfortunately for him, Argand was not one of the *ferblantiers*, who alone had the guild right to sell lamps in France; his lamp was pirated by one called Quinquet in France, and the patent was not protected in England, where he had gone to perfect its manufacture. He was forced to come to an arrangement both with Quinquet and with another distiller, called Lange, who improved the design of the glass funnel by narrowing the upper part. The Revolution abolished all industrial copyrights, but Quinquet continued successfully to manufacture the lamps in France, where they were named after him, thus receiving both the credit and the profit of the invention there, whilst firms like Boulton and Watt marketed the invention, using Argand's name, in England *[fig.22]*.

Figure 22.
A colza-oil (Argand) standard-lamp of bronze and ormulu, manufactured in England, *c*.1830–5.

Here one of the earliest uses that they were put to was illuminating shop-window displays for passers-by, to the delight of foreign visitors like Sophie von la Roche in 1786, for whom the displays were themselves a novelty. Argand lamps and Quinquets both ran on whale- or rape-seed oil (called colza oil, because it comes from a kind of kale) which, being heavy, was fed into the lamp from a reservoir above (throwing an inconvenient shadow) through a regulating valve; in 1800 the Carcel lamp, also named after its inventor, was patented, which used a clockwork pump to force up the oil from below, and in 1836 Franchot invented the Moderator lamp, which used a spiral spring to do the same. A novel feature of these lamps was that it was possible, by regulating the wick and the flow of oil, to control the intensity of light they emitted.

Argand and kindred lamps were a vast improvement on candles, but they did not displace these, because their brightness was an extravagance still found unnecessary (in 1846 an Argand lamp was found to give the light of, and cost roughly the same as, seven mould candles), and indeed harsh (Mrs Trollope tells approvingly of a party in Vienna in 1837 at which the contrast between the 'clear pure light of wax' candles illuminating the dining-rooms, and the lamps lighting the other rooms caused the host to substitute candles in these rooms too) and even bad for the eyes (Mme de Genlis, in her *Dictionnaire des étiquettes* (1818) ascribed the increase of neurasthenia and of spectacles amongst the young to the displacement of candles by lamps). In 1827 Antoine Caillot noted the use of wax candles instead of lamps as one of the marks of the old aristocracy. A lithograph by Karl Girardet in the *Magasin Pittoresque* of December 1847 showing a section through a Parisian *immeuble* at night not only shows the subtle gradations of light according to floor (and hence social station) and activity, but also shows that illumination was still almost entirely by candle *[fig.23]*.

The discovery of a fuel distilled from mineral oil, paraffin, in 1830 introduced a light fluid that fed the wick by capillary action alone, making the innumerable varieties of oil-lamp

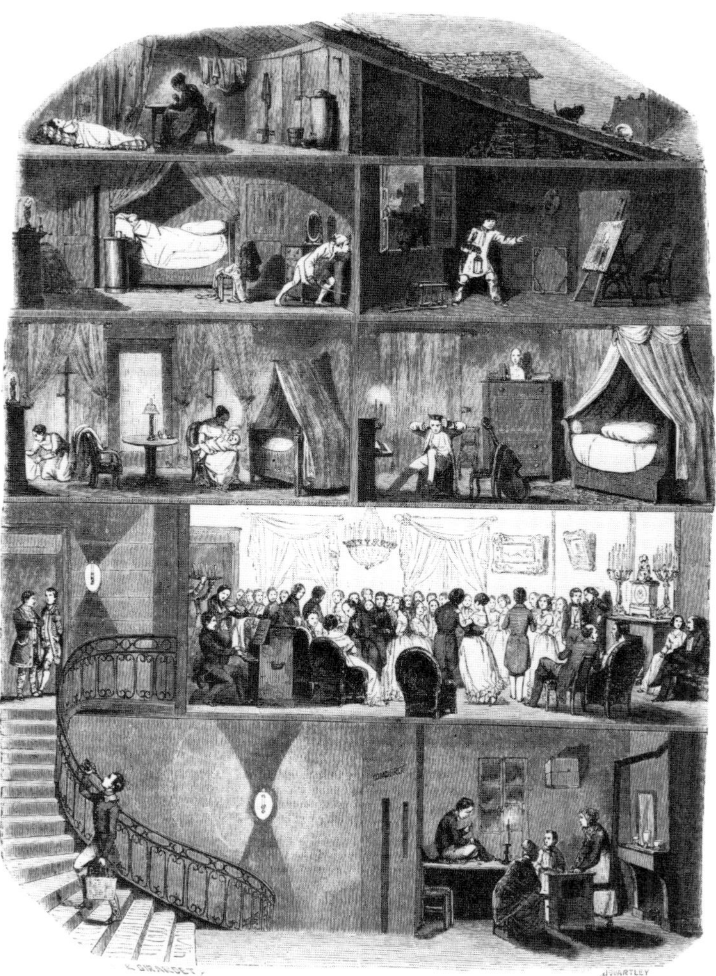

Figure 23. A section through a Parisian *immeuble*, showing the kinds of lighting appropriate to different classes and occasions. From a lithograph by J. Quartley after Karl Girardet in *Le Magasin Pittoresque*, December 1847.

obsolete, but it only became cheap and plentiful with Colonel Drake's discovery of petroleum oil in Pennsylvania in 1859, by which time paraffin lamps were forced to compete with the advances in lighting by gas and electricity, only becoming popular in remoter rural areas where public supplies of the latter were not available. Experiments in the production of gas had been made since the seventeenth century, and it had been used to light offices by Spedding in Whitehaven as early as 1765, and to light a factory by Boulton and Watt in Soho, Birmingham, in 1798. Gas street-lighting as we have seen, began in 1807. Experiments in the use of gas for domestic lighting had been made by Lord Dundonald at Culross Abbey in 1787, and by William Murdock at his house in Redruth in 1792–5. In 1818 the grounds, exterior, and great dragon chandeliers of the Banqueting and Music Rooms of the Royal Pavilion at Brighton were lit with coal gas. Nonetheless, the domestic use of gas lighting required first the creation of public companies to supply the gas, and then the perfection of burners by such inventions as Sugg's non-corroding steatite horizontal burner of 1858 to increase the ratio of light to heat, and his Christiana Governor burner of 1874 to keep the ration of gas supply to air steady, and above all, Baron von Welsbach's gas mantle, which produced incandescence without releasing soot, before it became a practical proposition in small enclosed interiors without adequate ventilation. Mrs Proudie complained at finding gas installed only in the kitchens and passages of the Bishop's Palace at Barchester, and promptly had it laid in the rest of the house, including chandeliers with twelve burners. Similarly, the earliest experiments in electric lighting were with the arc lamp, which was far too brilliant and impermanent for domestic use. Swan and Edison's virtually simultaneous invention of the vacuum light-bulb in 1878–80 produced the solution, and in December 1880 the armament manufacturer Sir William Armstrong installed the first domestic electric lighting in the house that Norman Shaw had recently built for him at Cragside *[fig.25]*. Another section of a house at

Figure 24. Homogeneity – the gas-lit Haussmannian *immeuble*. From a steel-engraving in *Le Magasin Pittoresque*, May 1891.

night from the *Magasin Pittoresque [fig.24]*, this time from May 1891, already shows a block of flats illuminated by electricity from top to toe – in Paris its installation was accelerated by the fears aroused by the burning-down of the Opéra-Comique in 1887. The notable feature of this section, by contrast with that of 1847, is the uniform lighting of the rooms (though this is also due to the greater social homogeneity of the post-Haussmannian *immeuble*, just as the emphasis on ceiling lighting reveals a straight switch from gas); only the luxuriousness of the light-fittings differs. Proust alludes to the novelty in *À l'ombre des jeunes filles en fleur*: 'Talking of eyesight, have you heard that the new house Mme Verdurin has just bought is to be lighted by electricity? . . . Even the bedrooms are to have electric lamps, with shades which will filter the light . . .' It is highly probable that the new ubiquity and intensity of light was a contributory factor in the disappearance of the overcluttered, dingy and dust-gathering Victorian interior.

Electricity at first met with some of the same snobbish resistance as Argand lamps had, and only gradually were its great advantages over gas appreciated: its safety (no naked flame), cleanliness (helping to make 'spring cleaning' a thing of the past), and, above all, flexibility: aptly-named flex made it once more possible to put standard-lamps and table-lamps where required. At the same time technical advances in electronics evolved special forms of light for particular purposes – neon lamps (invented in 1910) for street-lighting and advertisements *[pl.9]*; fluorescent lamps (shown 1935) for office-blocks, because of the great increase in interior floor-space that their low discharge of heat made possible; and spotlights for illuminating particular areas or objects. Dimmer switches recovered the adjustable intensity of the lamp. For, despite attempts at modernism, home users continue to prefer the atavistic reassurance of lamp-like lights muted by a shade.

7 Lampshades

Early photographs of Sir William Armstrong's house *[fig.25]*, and of the Great Hall of Stokesay Court (reputedly the first house in England to be designed for electric lighting, built by Thomas Harris in 1889) show naked bulbs hanging down; but from early on it was usual for electric lights (a 100-watt bulb gives almost one hundred times as much light as a wax candle) to be guarded from the eye by glass lamp-shades, like oil-lamps before them (whence the name). The first lampshades were designed less to filter the light than to keep the distraction of a bright light-source from one's eyes, especially when reading. The earliest seem to have been used on the

Figure 25. Lighting on the staircase in Cragside, Northumberland, installed *c.*1881 (photo 1891). Note the absence of shades, and the punty at the tip of the vacuum bulb.

rather brighter oil-lamps, like the metal shield on the raisable triple-spouted reading-lamp *[pl.7]*, and the earliest cylindrical lamp-shade to be reproduced, that on the lamp carried by Larmessin's Ferblantier. The transfer from lamp to candle is recorded in Pepy's gratitude to a friend in 1669, who 'hath beyond his promise, not only got me a candle-stick made after a form he remembers to have seen in Spain, for keeping the light from one's eyes, but hath got it done in silver very neat and designs to give it me'. In the late seventeenth and eighteenth centuries lamp-shades in a variety of materials became not uncommon, particularly those with reflecting white interiors, and soothing green exteriors, of ivory, ebony, or silk. The Princesse de Craon was reported by Horace Mann to use 'a *fanale* with a green shirt on'. Mme de Genlis records their use for reading as common before the Revolution, whilst Horace Walpole exulted in his sight recovering sufficiently to need neither 'spectacles nor candle-screen'. The multiple-candled late-eighteenth-century French *bouillotte* lamps (so-called from the fast-moving game played by their strong light) always had cylindrical shades. With the more powerful Argand, gas, and electricity lamps of the nineteenth century lamp-shades became a necessity, and the favourite material was clouded or patterned glass. In our times, as electricity has reduced distinctions between the light-fittings themselves, the lamp-shade (which is also necessary to deflect light downwards from overhead light-sources) has taken over as one of the chief elements of differentiation and decoration. Light-shades distinguish 'domestic' lighting from the pitiless glare of the lighting of work and public places, whilst transmitting messages about their users: from the large solid shades upon standard-lamps and table-lamps of the upper classes (who shun overhead lighting as a vulgarity, and revert to the separate pools of light of candle days) *[pl.12]*, to the Japanese paper globe lanterns adopted by the 'new Brahmins' in pursuit of conspicuous non-consumption; we all find a naked light-bulb as unseemly as the Victorians found an undraped chair-leg.

Bibliography

Whilst there are a number of books about the physical implements of lighting, such as candlesticks, the literature on the use and roles of lighting through the ages is astonishingly meagre, doubtless because the evidence is so scattered and hard to pursue systematically – a medieval document here, a picture there, or some passing mention in a letter, travel-book, or novel. There are, however, two masterly books on the subject: those by d'Allemagne, *Histoire du luminaire*, and O'Dea, *The Social History of Lighting*; my indebtedness to the latter in particular is immeasurable. I list below some of the works that were of the most use to me.

Henry-René d'Allemagne, *Histoire du luminaire*, Paris, 1891.

Donald M. Bailey, *Greek and Roman Pottery Lamps*, London, British Museum, 1972.

Reyner Banham, *The Architecture of the Well-tempered Environment*, London, 1969.

P. V. C. Baur, *The Excavations at Dura-Europos*. Final Report IV/3: *The Lamps,* New Haven, 1947.

Veronika Baur, *Kerzenleuchter aus Metall*, Munich, 1977.

E. S. de Beer, 'The early history of London street-lighting', *History*, N.S. XXV (1941), pp.311–24.

Die Beleuchtung in alter Zeit, exh. cat. Berlin, Märkisches Museum, 1928.

Ladislaus von Benesch, *Das Beleuchtungswesen*, Vienna, 1905.

Alan St H. Brock, *A History of Fireworks*, London, 1949.

C. N. Brown, *J W Swan and the invention of the Incandescent Electric Lamp*, London, Science Museum, 1978.

Reinhard Büll, *Das grosse Buch vom Wachs*, Munich, 1977.

Thomas Burke, *English Night Life*, London, 1941.

Elizabeth Burton, *The Elizabethans at Home*, London, 1958; *The Jacobeans at Home*, London, 1962; *The Georgians at Home*, London, 1967.

M. St Clare Byrne, *Elizabethan Life in Town and Country*, London, 1925.

J. C. Cox and A. Harvey, *English Church Furniture*, London, 1910, ch.X: 'The lights of a church'.
D. R. Dendy, *The use of lights in Christian worship*, Alcuin Club colls. XLI, London, 1959.
Dictionary of English Furniture, revised by Ralph Edwards, London, 1954.
Duhamel du Monceau, *Art du Chandelier*, Paris, 1762; *Art du Cirier*, Paris, 1762.
John Dummelow, *The Wax-Chandlers of London*, Chichester, 1973.
Alastair Duncan, *Art Nouveau and Art Deco Lighting*, London, 1978.
Otto von Falke and Erich Meyer, *Romanische Leuchter und Gefässe*, Berlin, 1935.
Daphne Foskett, 'Georgian domesticity in the North', *Country Life*, 27 November, 1975.
John Fowler and John Cornforth *English Decoration in the 18th Century*, London, 1974, ch.7: 'Lighting and heating'.
Larry Freeman, *Light on Old Lamps*, Watkins Glen, 1944.
Mark Girouard, *The Victorian Country House*, Oxford, 1971.
Henry Havard, *Dictionnaire de l'ameublement et de la décoration depuis la XIIIe siècle*, Paris, 1877–90.
Gabriel Henriot, *Encyclopédie du luminaire*, Paris, 1933–4.
Egon Hessling, *Documents de style Empire : le luminaire*, Paris, 1911.
Josef Holey, 'Die Kristalkronleuchter; seine Entstehung und Entwicklung', *Stifter Jahrbuch VIII* (1965), pp.7–36.
Richard Hubbard Howland, *The Athenian Agora*, vol.IV: *Greek Lamps and their survivals*, Princeton, 1958.
G. Janneau, *Le luminaire de l'antiquité au XIXe siècle*, Paris, 1934.
Kurt Jarmuth, *Lichter leuchten im Abendland*, Brunswick, 1967.
Lampe, Leuchter, Laterne, exh.cat. Munich, Neue Sammlung, 1964–5.
'Let there be light', exh.cat., Hartford, Wadsworth Atheneum, 1964.
Dorothy Marshall, *Johnson's England*, Oxford, 1933.
Erich Meyer, 'Der gotische Kronleuchter in Stans. Ein Beitrag zur Geschichte der Dinanderie', *Festschrift Hans Hahnloser*, Basel, 1961.
Ronald F. Michaelis, *Old Domestic Base-Metal Candlesticks from*

the 13th to the 19th century, Produced in Bronze, Brass, Paktong and Pewter, Woodbridge, 1978.

Randall Monier-Williams, *The Tallow-Chandlers of London*, London, 1970–3.

William T. O'Dea, *The Social History of Lighting*, London, 1958; *Making Fire : Wood friction, tinder boxes, matches*, London, Science Museum, 1964; *Lighting, 1 : Early oil lamps, candles*, London, Science Museum, 1966; *Lighting, 2 : Gas, mineral oil, electricity*, London, Science Museum, 1967; *Lighting, 3 : Other than in the home*, London, Science Museum, 1970.

Paulys Real-Encyclopädie der classischen Altertumswissenschaft, Stuttgart (s.vv. *candela, Fackel, funale, lanterna, lucerna*).

Anthony F. Radcliffe, 'Bronze Oil Lamps by Riccio', *Victoria & Albert Museum Yearbook*, no.3 (1972), pp.28–58.

F. W. Robins, *The Story of the Lamp (and the Candle)*, London, 1939.

L. F. Salzman, *English Life in the Middle Ages*, Oxford, 1926.

Still the Candle Burns, privately printed for Price's Patent Candle Company, Battersea, 1972.

W. G. Mackay Thomas, 'Old English candlesticks and their Venetian prototypes', *Burlington Magazine* LXXX (1942), pp.145–51.

Leroy Thwing, *Flickering Flames*, London, 1959.

Sigrid Wechssler-Kümmel, *Schöne Lampen, Leuchter und Laternen*, Heidelberg, 1962.

Stanley Wells, *English Candlesticks before 1600*, London, 1954; *Period Lighting*, London, 1975.

Gerhard Wietek, *Altes Gerät für Feuer und Licht*, Oldenburg, 1964.

Geoffrey Wills, *Candlesticks*, Newton Abbot, 1974.

Adam Winter, 'Brennende römische Tonlampen', Saalburg Jahrbuch, 1955.

Thomas Wright, *The Homes of Other Days*, London, 1871.

Index

Printed in England for Her Majesty's Stationery Office by McCorquodale Printers Ltd, London
Dd 696345 C50